THE ATOM
WHAT AM I REALLY?

WRITTEN BY
TERRESSA BOYKIN

ILLUSTRATED BY
AARON PERSH

des&desi ℠

ISBN-13: 978-1-945751-01-1

Library of Congress Cataloging-in-Publication Data
2016952120

Boykin, Terressa
des&desi
The Atom: What Am I Really?

THIS BOOK IS DEDICATED
TO DESMEN AND DESIREE

WHAT AM I **REALLY**, AS I SEARCH NEAR AND FAR?

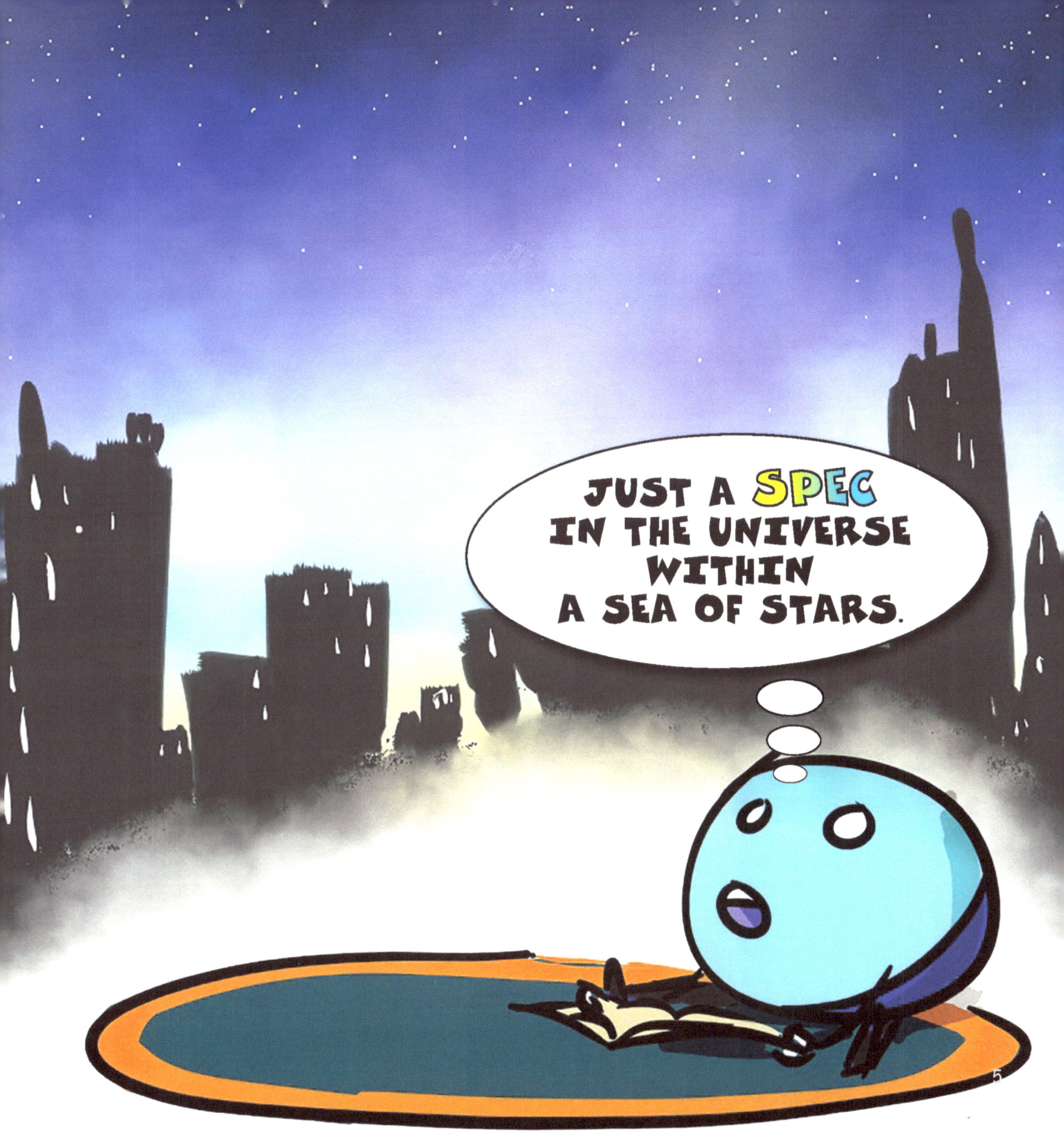
JUST A SPEC IN THE UNIVERSE WITHIN A SEA OF STARS.

AS I SIT IN MY CLASS ALONE AND FAINTHEARTED

I PONDER WHAT I AM AND HOW I WAS STARTED.

MY TEACHER SAID THAT READING CAN HELP ANSWER OUR QUESTIONS. READING GIVES US INFORMATION AND LOTS OF SUGGESTIONS.

SO PERHAPS, IF I START READING MORE BOOKS ON THIS SHELF
I'LL LEARN THINGS ABOUT ME, YES, MORE ABOUT MYSELF.

PLEASE LET ME EXPLAIN.

YOU ARE AN ATOM

IT IS ALL VERY PLAIN.

THE ATOM MAKES MATTER ALL IT CAN BE.

ONE OF THE TINIEST OBJECTS MAKES THE MATTER YOU SEE.

ALTHOUGH YOU CAN'T SEE THEM ATOMS ARE ALL AROUND.

THEY ARE THE BLOCKS THAT BUILD THINGS FROM THE SKY TO THE GROUND.

ATOMS CAN MOVE AROUND IN ONE PLACE.

OR **FLY**
LIKE OBJECTS
FLOATING
IN SPACE.

ATOMS LOOK LIKE TOY BALLS
THEY ARE SPHERICAL IN SHAPE.

MUCH LIKE AN ORANGE,
A MELON OR A GRAPE.

BAM!!!

ATOMS DON'T DRIBBLE WHEN THEY BOUNCE.

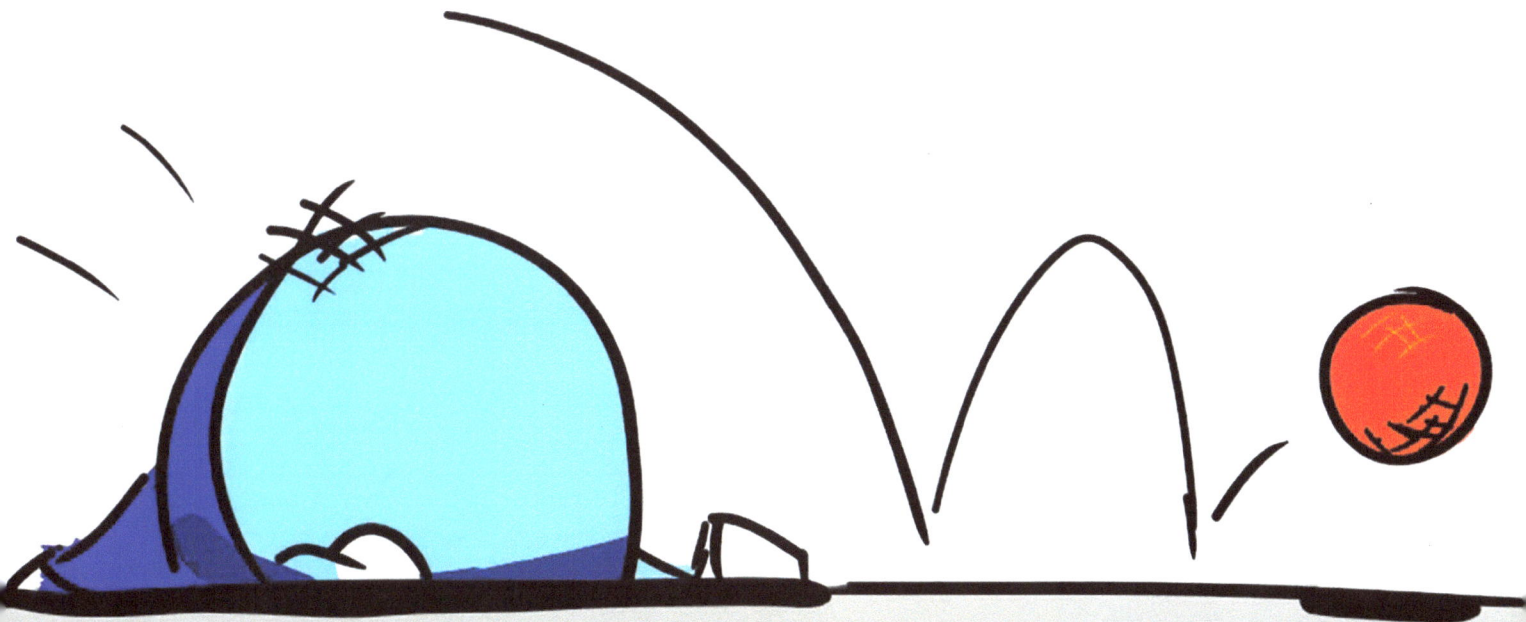

AND,
WEIGH LESS
THAN
FEATHERS...

YES, LESS THAN AN OUNCE.

ATOMS COME IN MANY GROUPS AND IN ALL DIFFERENT SIZES.

THEY HAVE MANY STRANGE NAMES AND EVEN SILLIER DISGUISES.

LIKE FRIENDS HOLDING HANDS, EACH ATOM IS ATTACHED.

AND LIKE TO STICK TOGETHER, BY THEIR BONDS THEY ARE LATCHED.

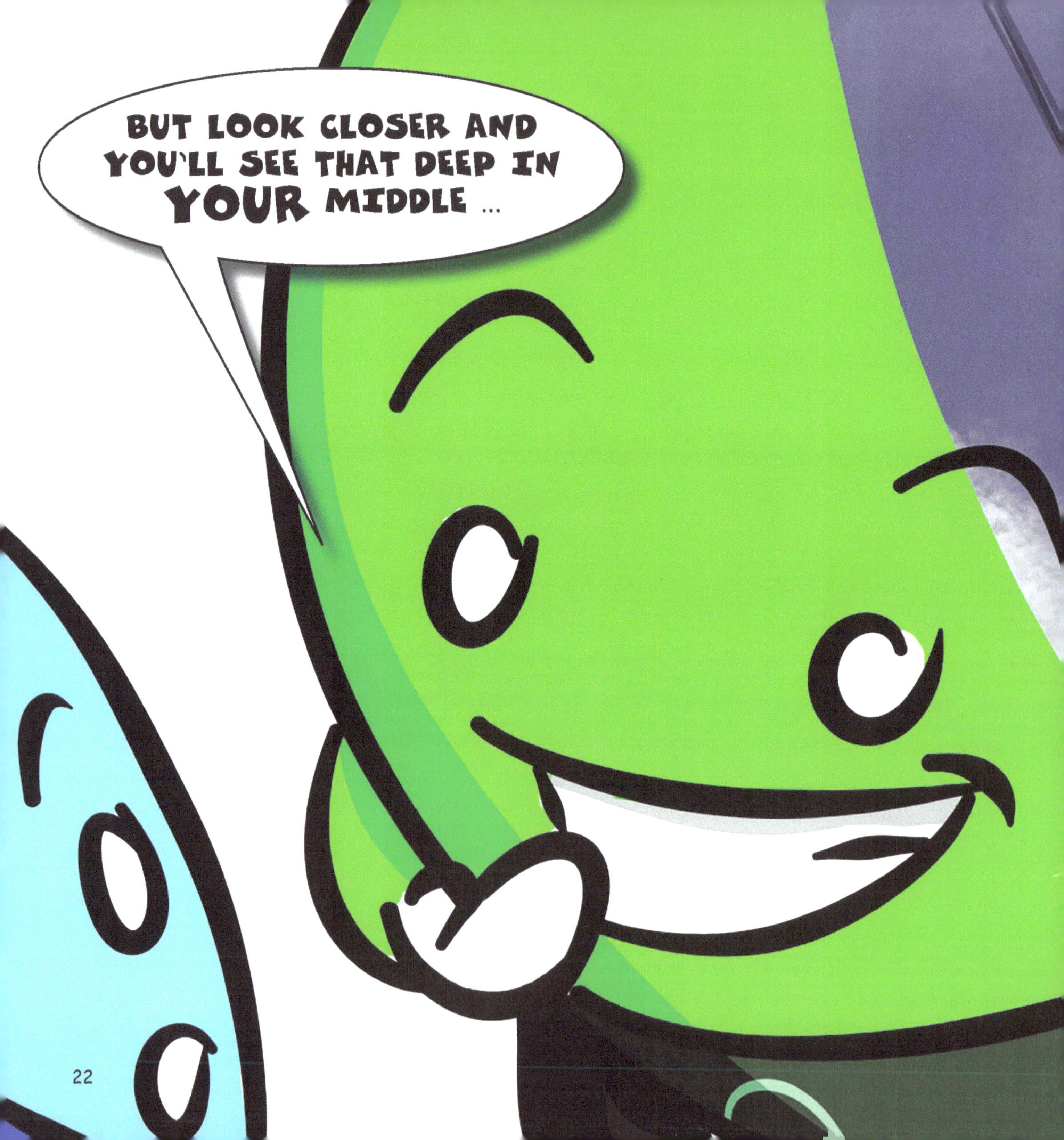
BUT LOOK CLOSER AND YOU'LL SEE THAT DEEP IN YOUR MIDDLE ...

... IS A CENTER CALLED A **NUCLEUS** THAT'S TINY AND LITTLE.

INSIDE THIS NUCLEUS THERE IS A GROUP OF SPHERES

CALLED **PROTONS** AND **NEUTRONS** HELD TIGHTLY AND DEAR.

SURROUNDING THE NUCLEUS SPINS AN **ELECTRON** IN MOTION

CIRCLING AROUND IT LIKE FISH IN THE OCEAN.

THIS ELECTRON KEEPS MOVING LIKE A BALL ON A STRING.

AROUND THE NUCLEUS IT SPINS MAKING A SPHERICAL RING.

EACH PARTICLE HAS A CHARGE OF ELECTRICAL DUST.

THE ELECTRON HAS A NEGATIVE CHARGE
AND THE PROTON A PLUS.

THE NEUTRON HAS NO CHARGE AND IS NEUTRAL IN FACT.

WHEN ALL ARE TOGETHER THERE IS A BALANCING ACT.

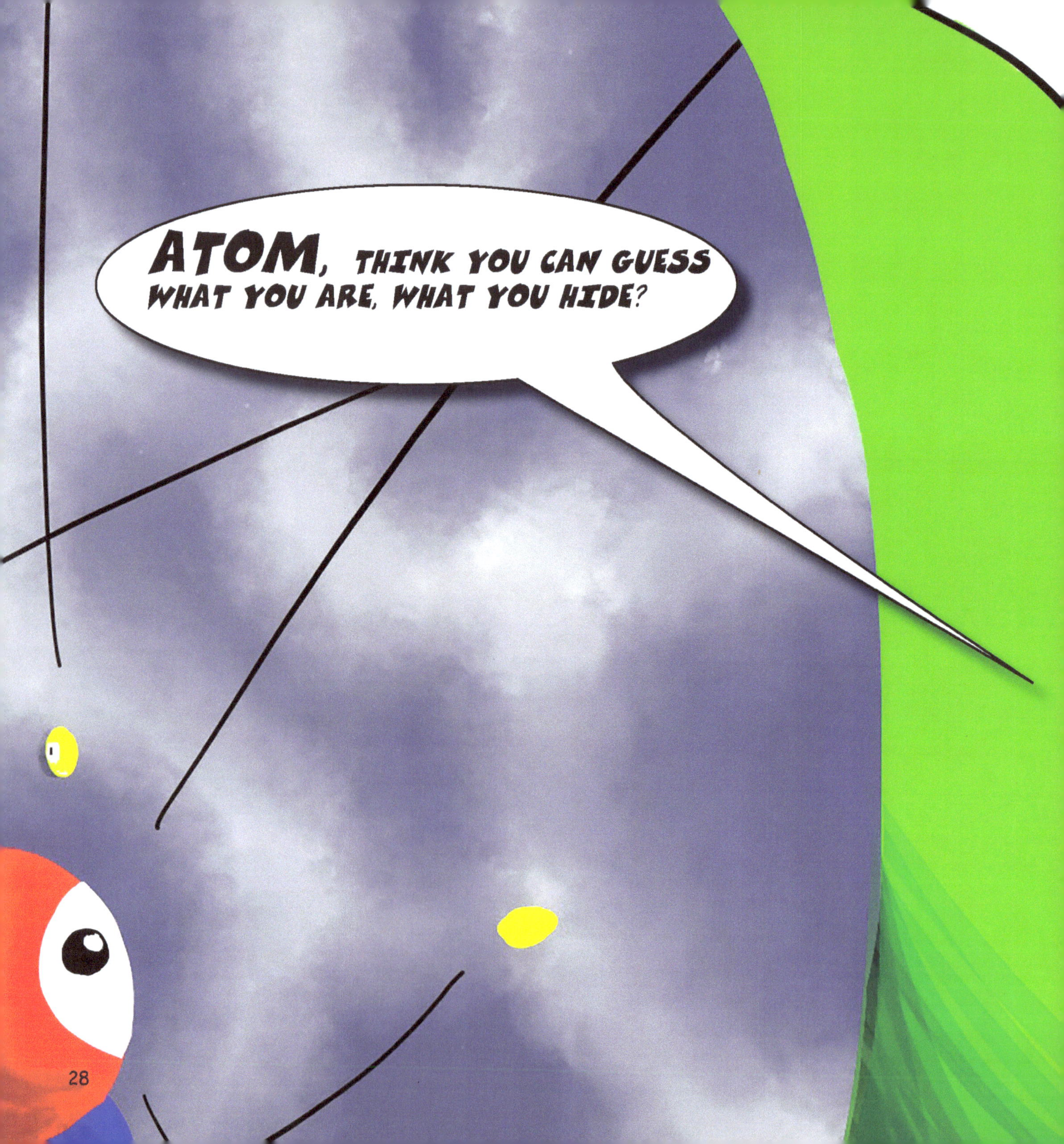
ATOM, THINK YOU CAN GUESS WHAT YOU ARE, WHAT YOU HIDE?

NO NEUTRON, ONE PROTON, AND ONE ELECTRON INSIDE.

29

YOU WERE THE FIRST OF THE ELEMENTS AND WERE USED TO MAKE STARS.

YOUR NAME MEANS FORMING NOW—DO YOU KNOW WHAT YOU ARE?

YOU ARE THE FIRST, ATOM, THE FIRST ON THE CHART.

IN FACT, LITTLE BUDDY YOU WERE HERE FROM THE START.

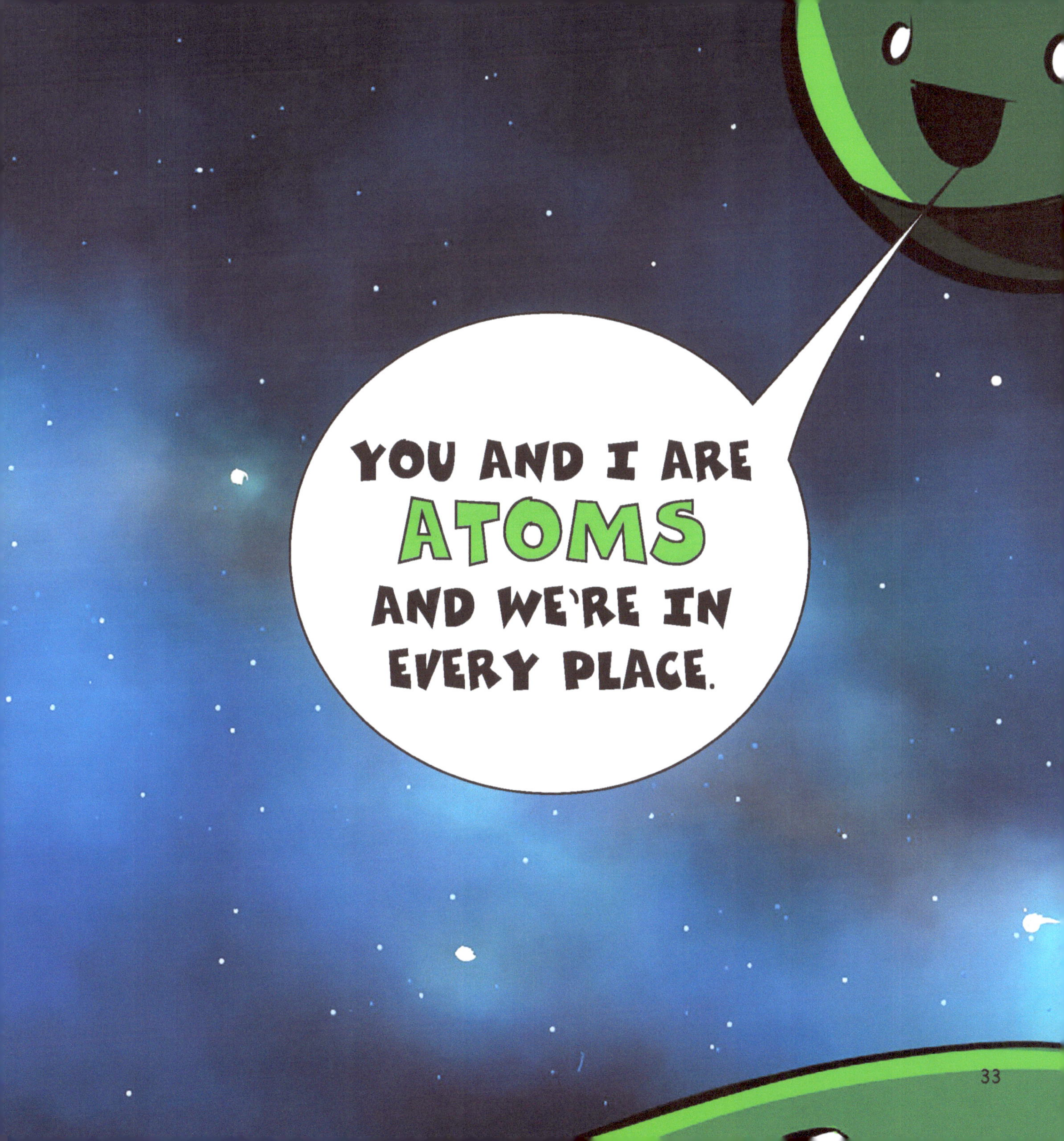
YOU AND I ARE ATOMS AND WE'RE IN EVERY PLACE.

ABOUT THE AUTHOR

TERRESSA BOYKIN HOLDS A MASTERS OF SCIENCE IN ORGANIC AND KINETIC CHEMISTRY AND HAS WORKED IN THE FIELD OF POLYMER CHEMISTRY AND CHEMICAL ENGINEERING FOR OVER TWENTY-EIGHT YEARS.

IN ADDITION TO VOLUNTEERING IN HER COMMUNITY AS A LOCAL AND REGIONAL SCIENCE FAIR JUDGE, SHE IS A MEMBER OF THE AMERICAN CHEMICAL SOCIETY'S COACHING PROGRAM, THE EDUCATION AND STEM COMMITTEE FOR HER LOCAL CHAMBER OF COMMERCE, AND IS THE FOUNDER AND CEO OF DES&DESI CHEMISTRY CLASSES FOR YOUNG CHILDREN. SHE IS THE AUTHOR OF THE BOOK SERIES DES&DESI CHEMISTRY_ IN MY DREAMS.

WWW.DESNDESI.COM

ABOUT THE ILLUSTRATOR

AARON PERSH HAS BEEN DRAWING HIS ENTIRE LIFE AND COULDN'T BE MORE EXCITED TO ILLUSTRATE HIS FIRST BOOK. STARTING WITH DOODLES IN HIS SCHOOL NOTES AND COMICS ON LINED-PAPER, HE DID NOT CONSIDER ANYTHING ELSE OTHER THAN SKETCHING WITH A MECHANICAL PENCIL. NOW, AS AN ILLUSTRATION STUDENT AT THE RINGLING COLLEGE OF ART AND DESIGN HE COULD NOT BE HAPPIER TO PURSUE HIS DREAM TO CREATE ART.

MY INSPIRATION

DESMEN A. BOYKIN IS A MIDDLE SCHOOL STUDENT WHO HAS A PASSION FOR SCIENCE AND A TALENT FOR CREATIVE DRAWING. HE HOPES TO BE AN ENGINEER ONE DAY.

DESIREE M. MOSLEY IS AN LPN AND CURRENTLY AN RN NURSING STUDENT WHO HAS A GIFT AND A PASSION FOR HELPING OTHERS.

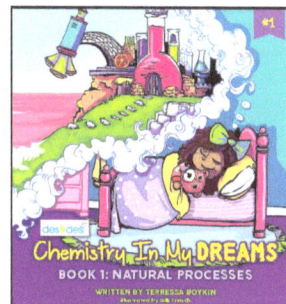

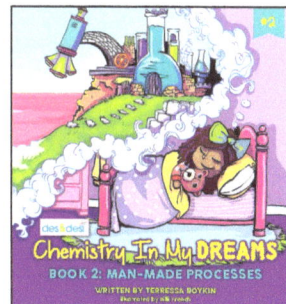

www.ingramcontent.com/pod-product-compliance
Lightning Source LLC
Chambersburg PA
CBHW052045190326
41520CB00002BA/189